UNANIMOUS ANONYMOUS VIII

Lincoln Plant after 1985 plus ESP

This is a story of what happened after 1984 and my chemical imbalance that gave me a touch of ESP. I believe it is like what some soldiers experience in war when sometimes they acquire an extra sense to protect themselves.

This is a more detailed story in continuation of the UNANIMOUS ANONYMOUS SERIES. It is a compilation of extra information acquired through ESP and my own intuition which became quite keen.

My book on Cernik Work History explains my work in the 1982, 1983, & 1984 time periods. The Lincoln Plant management thought they were different and tried to control with a microscope their interaction with the corporate headquarters in Akron. Akron corporate forced the Lincoln plant in doing things their management staff was resistant to. Thus, in my first book I made them their own corporation because the super powerful Local management and Union thought they could do anything they wanted including violating corporate policy. They thought they were gods doing anything they wanted and sold Akron Corporate with an expensive and waste-full bill of goods and corporate fell for Lincoln Plant misjudgments, misinterpretations.

Skip Westmoreland was a very powerful man locally like the blind leading the blind. Skip was always fighting battles with corporate management fighting for the Lincoln plant to survive. My experience was that a non-profitable division was usually sold and the name contracted to the buyer for a period of five years to help the purchaser sell products in name only. This corporate strategy was used in many division sales that I witnessed.

Skip had his own strategy, he would do whatever was in his power to keep jobs in the Lincoln Plant to keep his friends and cronies working their until they could get retirement and draw their retirement pensions. Skip was put in charge of building a cut edge process plant in Taiwan. They spent millions of dollars getting the Taiwanese plant operational only to close it through Westmoreland political maneuverings. He told Akron after plant was in existence that belts were not selling because the high humidity caused the belt to squeak on drives, any intelligent rubber engineer new the problem was material related and required a compound material change to eliminate squeak. Skip however got Akron to close the plant and Lincoln jobs survived.

When I came back to Lincoln in 1985 Skip was the product development manager. Of course, he didn't have an engineering degree and I wasn't sure what he had but he had a lot of friends in the Lincoln Plant. He would promote the good old boys whether they were qualified or not.

After coming back to Lincoln, he was told by Akron to promote me but he didn't want to do it. He believed in an antiquated world war II command and control structure and was afraid of losing control if he made change to the management structure. Akron Advanced Concepts department taught me how to cut red tape. One day I wanted to talk to him and went into his office. He got up and slammed the door behind me and yelled at me why did I come back to Lincoln? I gave him a partial reason and after that he continued to try and manipulate my personal life, even hiring a secretary and telling her she was supposed to Marry me. Sid fix was the manager under him that I worked for and it was Sid that called me in New Bedford and wanted me to come back to Lincoln and take a job in the Transmission Department. He had ulterior motives and operated on his own agenda.

One time Skip was walking by and I read his mind. Skip wanted to see what made me tick. I never let him figure it out. He sent me to HR Tom Booths office and Tom told me Skip was a very busy man and didn't have time to talk to me with all the people reporting to him. I thought to myself that that amounted to a hello in the hallway and that is all Skip got after that.

Sid expected me to come in his office and talk to him like I brown nosed him when I worked for him in the old Advanced Technology department prior to Advanced Concepts and the Advanced Technology department getting dissolved. They didn't realize that if they wanted to talk to me all they would have had to do was call me in their office and ask questions. I would have answered them instead they expected me to come in there and voluntarily spill my guts.

There was a start of a power struggle in the department. Del Stork and Doug Gilg were in the robotics department. A department I felt was a waste money. They had Hose trim people take a hose put in a fixture so a robot could pick it up and then trim the end off and put in a box. The worker had the hose in his hand why waste money on robotics other than eliminate some cut fingers careless operators would let happen. Most of the people had a high school education and that is all they knew how to do. It gave them a high paying Union job which our Advanced Concepts department cost studies indicated why the plants had trouble making a profit. Stork and Gilg also tried to get robots for hose pulling job. Which I knew was a waste of millions of dollars. Later when they moved department to a Mexico location sometimes it would take two Mexicans to do what one American could do in the plant. The engineer in charge of developing the cell knew it wouldn't work and quit Goodyear before the Hose pull waste of money cell was completed.

Skip sold Clay Orme a bill of goods saying their automation would save money and a modernization program would save jobs. Clay went before

Nebraska legislature to get tax incentives for a modernization program and all it did was scrap old equipment and reorganize product flow through an already archaic plant which didn't help much but waste money.

With Westmoreland being assigned the Taiwanese plant project that left an opening for his department manager replacement. Sid got me back to Lincoln to help me get him the job he knew Skip would be reassigned and Del Stork because of his facade of modernization also made him a management candidate. Because of my Akron experience I was a threat to them because I would send emails to Akron Corporate telling them what I thought. Skip decided he had to discredit me so he had to come up with a reason to tell Akron to ignore me. He forced me to see what they so called the plant shrink who was nothing more than an ex-UNL HR person by the name of John Holmes. He had no psychiatric degree. I had a lot of conversations with him and he knew the real reason Skip wanted me to see him, Skips so called plant Shrink. One time I went home and took a few days' off because I was furious with those people. I called John up and left a message that I was going to sue them. When I went back to work a few days later John had quit. I knew what Skip was trying to do and I said to myself once I was going to quit my Job, get my law degree and make a living suing them and the people that they were. I also in anger told Mike Hotovy that this may end up in court but I would win. There were too many of them they thought there was safety in numbers. So, in 1987 I decided to wait till they all retired and make them look bad which the Unanimous Anonymous series is an attempt at doing.

Also, about this time, Sid Fix told the law enforcement agencies I was selling information to the competition which was a blatant lie, but it did get the FBI involved Which I could prove by the year 1992. Also, the segment belt project was a revolutionary idea and the Union was afraid of it because they might lose some jobs. A Mechanic Union Stewart out of the shop by the name of Sullivan had a big burly hose puller waiting for me when I was walking in the gate one morning. Sullivan told the hose puller their he is as I walked in and the hose puller proceeded to follow me into the plant in an attempt to intimidate me. Sid Fix got his way and killed the segment belt project by transferring me to synchronous belts working for Gene

Zitek. The same man that refused to interview with Barry Long and an Advanced Concept job.

This all depressed me and I saw Dr. Leslie Spry a kidney doctor specialist who prescribed Tofranil PM for depression. If you look at the side effects of this medication I experienced all of them. I however not realizing took the medication for six months and it made my chemical balance worse. It took a Dr J Bowman Bastani another six months to get the chemical imbalance in somewhat of a control. However, from what Bastani was saying I knew Skip Westmoreland was or someone from the plant was talking to Bastani and he had me on the wrong medication based on plant management lies. He had me so doped up you could have punched me in the face and I would have said that is nice, do you want to do it again. I went to a Dr Blodig in Omaha to get a second opinion. That only made things worse because by that time Del Stork was the new Department manager. I was working for Gene Zitek and told know one why I was seeing a Doctor. From Jason Kline, I found out Del Stork was trying to get information out of Blodig and Blodig was giving him a bunch of lies that Stork repeated around the department and telling Jason Kline. The Lincoln Plant management violated HIPPA laws and it was very difficult to get a doctor to prescribe the right medication for my Chemical imbalance. It wasn't till 1992 when I did an end around and saw Dr Osborne and finally got the correct medication that I still am on today. The New Bedford project Skip Westmoreland and Del Stork Messed with my medical and personal life which was none of their business other than their thirst for management power. Dr Blodig was associated with a for profit psychiatric hospital in Omaha and all those doctors were interested in doing was find patients to put in the hospital so the hospital could make a profit. The hospital went out of business a few years later. But not before writing a pack of lies about me. This however I knew how to keep out of the wrong hands. They didn't want to get involved with a lawsuit with Goodyear Lincoln Plant management so it was a bunch of fabricated lies to make me look bad. Stork won right up to the time he got fired.

A Director in Akron helped me. He knew the Lincoln Plant was doomed as a Plant and let me get out in 2003 only to sell the Division and plant in 2007.

My first fiction novel in 2004 and 2005 stirred up the pot and made Lincoln Plant management, Del Stork, Skip Westmoreland and Clay Orme Furious only getting the Nebraska Department of Road's Curt Mueting, State Patrol, and FBI involved since they could no longer do anything to me in the Lincoln Plant. The Lincoln Plants days with that company was numbered.

Now owned by a new company. From What I assume is that the new company changed all the material ingredients to better compounds and automating the antiquated belt building equipment with new equipment. It however now must be their way and not the old Plant management way. I hope those people learned a lesson. But there is no solution for stupid people doing stupid things.

Thus, so much for my venting of that Plant's management messing up my life. They did a lot more too numerous to mention in this novel. I may end up in court with these people if they have not died yet and I like to keep a lot of facts in reserve.

This section consists of a research paper I wrote while getting my Masters of Arts in Management Degree at Doane University graduating with a 3.998 GPA in December 2011.

How real is Telepathy and ESP?

Review of Literature

Leonard G Cernik

Doane University

Masters of Arts in Management

MAM program

Foundations of Research

BUSINESS Course

May 18, 2009

How real is Telepathy and ESP?

Introduction

The world is searching to find out if telepathy and ESP (Extra Sensory Perception) is real. Based upon the amount of research that has been conducted in this area of interest there is a good possibility telepathy is a real phenomenon. Additional research in other similar areas of study indicated differing degrees of confirmation of telepathy and ESP. Some studies showed indications that telepathy and ESP is more than mere chance. Other studies indicate that based on statistical probability, the probability of telepathy and ESP's existence was equal to that of chance.

Brown and Sheldrake (2001) surveyed two hundred people. The researchers found that three quarters (seventy-eight percent) of the people surveyed said that they telephoned someone who said they were just thinking about telephoning them. Besides this response to telephone calls, forty-five per cent of those surveyed said they experienced other situations that seemed to involve telepathy. A quarter of these people (twenty-two percent) said that their seemingly telepathic experiences happen often. These surveys were conducted in California and indicated there is a good percentage of chance that telepathy and ESP does exist. If these surveys are reliable, they provide evidence to support the reality of telepathy and ESP.

Sheldrake and Smart (2003) conducted more experiments that indicated a high percentage of chance that telepathy and ESP may occur. They conducted two hundred seventy-one experiments, all videotaped to eliminate the possibility of cheating. Participants in telephony videotaped experiments gave significant, repeatable results. When callers were known, there was a sixty-one percent success rate in predicting the caller and a forty-one percent success rate in prediction among unfamiliar callers. The experiments' results are additional evidence to support the reality of telepathy and ESP. These experiments occurred in the everyday environment of anticipation and prediction of contacts by phone calling. Distance between telephone callers was not a factor and more than mere chance.

Sheldrake and Smart in the press (July 2003), described the results of five hundred seventy trials involving sixty-three participants. Findings showed that the average success rate was forty percent, which is hugely significant statistically. Distance between callers was not a factor in that during these trials the callers were miles apart, and in some cases thousands of miles apart. The results would imply that the participants' above-chance success rate was a result of telepathy from the callers.

Smith (2005) discusses Sheldrake's theory. Sheldrake believes such phenomena are examples of 'seventh sense'. Sheldrake used the term 'seventh sense' since the more traditional 'sixth sense' is now more commonly used in the scientific community to refer to certain sensory activity found in the animal kingdom but not among humans. There has also been research indicating that owners and their pets have telepathic connections in cases where the pet knows when the owner is coming home, even at large distances between the owner and pet. Sheldrake has said that in research by Sir Rudolph Peters, a mentally retarded child sensed things through his mother, even when miles apart. The researcher theorized that this may be due to the intensity of the bond between mother and child. This bond may have developed because-it was necessary for the child's survival.

This literature review will focus on areas confirming or contradicting that telepathy and ESP are real phenomena. In the end, it will present conclusions on information about how real is telepathy and ESP. This review will discuss the following themes: the reality of telepathy, the relationship to ESP, the Psi phenomena, brain wave patterns that can be measured and read with sensors and MRI, and Ganzfeld experiments, which are experiments that measure ESP among test subjects.

Reality of Telepathy

Brown and Sheldrake (2001) conducted a survey in California on the anticipation of telephone calls. Two hundred random people were surveyed in Santa Cruz County, California. Data was collected by means of telephone surveys. Two hundred people participated. (Seventy-eight men and one hundred twenty-two women) Seventy-eight percent telephoned someone who said they were thinking about them. In the survey forty-seven percent, also new who was calling them before answering the phone. Sixty-eight percent said they had thought of someone who later telephoned that day. Some of these could have chance experiences, and others might have depended on telepathy. Two previous surveys in London and Greater Manchester England indicate this experience happens quite frequently.

The surveys are just another means of evaluating Telepathy.

Sheldrake and Smart (2003) videotaped experiments on telephone telepathy. Participants were videotaped during experimental period using four potential callers with one selected randomly. Two hundred seventy-one trials were conducted. The actual experiment used telephone anticipation instead of just survey. Random chance would indicate twenty five percent correct responses. The study of two hundred seventy-one total trials indicated forty-five percent correct guesses. With familiar callers, there was a sixty-one percent success rate. Videotaping trials reduced chance of cheating, if not making it virtually impossible.

There was a research report rebuttal written to the Editor (2003) of the journal that commented on Rupert Sheldrake's and Pam Smart's paper, "Videotaped experiments on telephone telepathy" The letter to the editor stated that data presented from experiments on telephone telepathy may be in error. In one of the specific analyses Sheldrake reported a significantly different hit rate dependent on whether the subjects were called by familiar or unfamiliar callers. The differences are reported in report to be "very significant statistically" and according to Sheldrake the significant difference "supports an interpretation in terms of telepathy". The letter to the editor stated that they disagreed with the analysis and the conclusion. They claimed that Sheldrake's conclusion was based on a statistically unsatisfactory procedure. The letter to the editor did comment that their critique applied only to the comparison of familiar and unfamiliar callers. The overall hit rate of the entire experiment remains untouched by their reanalysis.

Sheldrake and Smart (July 2003) described the results of five hundred seventy trials involving sixty-three participants. The participants were recruited through advertisements in a part time work section of newspapers or through a recruitment web site. These trials were a confirmation test for earlier studies. The significance is that these trials further documented the reality of telepathy. The findings were reported in correct determination of the random phone caller chosen between four potential callers. The overall success rate in caller determination was forty percent. This was hugely significant since it was above the twenty-five percent success rate expected based upon chance or random guessing. The emotional closeness of callers seemed to be more important than the physical proximity.

Smith (2005) wrote an article on telepathy research. Smith commented that far from being a cranky relic of a pre-Enlightenment dark age, the belief in telepathy would seem to be confirmed by contemporary science. Smith further commented in that telepathy might even help secure the planet's survival. Smith puts together a good summary on telepathic research and insights into Sheldrake's work and Sheldrake's professor at Cambridge. This article lists Sheldrake's web site www.sheldrake.org. This general presentation on topics related to the reality of telepathy was a good overview, providing research direction in finding proof for the reality of telepathy.

Vassy (2004) conducted a study of telepathy by classical conditioning. Experiments were conducted detecting telepathy by classical conditioning using a mild electric shock. The stimulus was a telepathic message and the unconditional response was the sudden rise in skin conductance. Fifty sessions of ten electric shocks were conducted, using conditioning assuming telepathy involves yet unidentified physiological process in the brain. Earlier studies were conducted in 1997 and 1978. There were significantly positive results in fifty runs. A replication study showed that the first experiment could not be replicated, thus raising questions about the study

Skeptic (2001) magazine showed results of a survey conducted on June 8, 2001 by the Gallup news service poll showing paranormal beliefs are on the rise. One thousand twelve adults eighteen years old or older were surveyed, conducted May 10-14, 2001 with a ninety-five percent confidence level. Eighteen to twenty-nine year olds are more likely to believe in telepathy and clairvoyance. People age thirty and older had different beliefs. Religious beliefs were a factor. Survey results showed that in telepathy thirty-six percent believed, twenty-six percent were not sure, and thirty-five percent don't believe. In two surveys on the belief in telepathy results showed that thirty-six percent believed in 2001, in a survey in 1990 thirty-six percent, these results show there was a zero percent change in between the two studies. Besides telepathy, other paranormal beliefs were studied, but belief in telepathy had not changed over time.

Sheldrake and Smart (2005) investigated possible telepathic communication in connection with e-mails. Trials were conducted with four potential e-mailers, one of whom was selected at random by the experimenter. One minute before a prearranged time at which the e-mail was to be sent, the participant guessed who would send it. Fifty participants (twenty-nine women and twenty-one men) were recruited through an employment web site. Five hundred fifty-two trials were conducted. The results of two hundred thirty-five guesses (forty-three percent) were hits. This is significantly above the chance expectation of Twenty-five percent.

Further tests with five participants (four women, one man, and ages sixteen to twenty-nine) were videotaped continuously. On the filmed trials, the sixty-four hits of one hundred thirty-seven trials or forty-seven percent were significantly above chance that expected by chance

Sheldrake, Godwin, and Rockell (2004) carried out experiments in an attempt to replicate the telephone telepathy phenomenon for a television show called "Are You Telepathic?" made by 20/20 Productions and broadcast in the UK on Channel Five Television on June 19, 2003. The participant and her four callers were Sisters, who had for years worked together in a girl band, the Nolan Sisters, popular in the UK in the 1980s. In most of the trials, the callers were in different locations from each other and were not filmed... However in one previous experiment, all four callers were in the same location, and the callers as well as the participant were filmed continuously. The test, carried out in Wakefield, Yorkshire, placed the participant one point five km away from the four callers. One of the sisters guessed correctly in eight out of seventeen trials (forty-seven percent); Sheldrake concludes that the results support the hypothesis of telepathic communication. Of course, further replications will be needed, and in future tests it would be desirable to rule out the possibility of cheating using mobile phones.

Wassner (2005) presented five stories about twins that have similar thoughts giving some proof to the reality of telepathy. The article stated that not all identical twins share a connection. This paper presents an example of five stories in which twins seem to be connected mentally in various ways. This is a general presentation of stories about identical twins providing examples where twins seem to have a connection that cannot be explained by today's societies view of what would be classed as a normal relationship.

Sheldrake (2006) used a telephone survey method to question nursing mothers about their reactions to when their baby needed them. Some nursing mothers claimed that when they are away from their baby they often knew when their baby needed them because their milk lets down. Some mothers were convinced that this response is telepathic. Sheldrake, in order to find out more about this phenomenon, surveyed one hundred mothers who had recently had babies and asked a series of questions about their experiences when breastfeeding. Sixty-two percent had experienced milk let-down when away from their babies. Sixteen percent had noticed that this seemed to coincide with their baby needing them. Most of these women had breastfed their babies for more than six months. In addition, three women said they had felt there was something wrong with their baby when they were away from home, and found that it was indeed in distress because of a fall or other accident. Five women commented that they often woke up shortly before their baby needed them in the night. These surveys used babies and mothers in another way of evaluating telepathy. Apart from the phenomenon of milk let-down, in question ten of all mothers in this survey were asked "While you were away from your baby, have you ever had a strong sense that your baby needs you, other than through your milk letting down?" Thirty-one out of one hundred answered yes to this question. Most of these women said in response to a previous question that this may be a matter of general anxiety or intuition. Several women commented that their anxiety was more related to their own worries than to their baby's needs. However, three women reported that they had felt something was wrong with their baby and either telephoned home or went to the baby and found that it was in distress because of a fall or other accident. Five women commented that they often woke up shortly before the baby needed them in the night.

Sheldrake (2005) presented various views based upon different people staring at other people or sensing that they were being stared at. Sheldrake presented different theories of how non-visual staring can mentally be detected. Sheldrake theorized that features of the perceptual fields may have implications for the understanding of morphic fields. The morphic fields may be related to patterns of activity in the electromagnetic fields of the brain but may be also separable from other electromagnetic fields. The brain can project virtual images in three dimensions with the basis of these patterns being electromag¬netic activity. The morphic fields may interact with patterns of electromagnetic activity not only through the eyes but elsewhere in the in the brain. This electromagnetic activity in an individual brain may be picked up by other people.

Sheldrake and Smart (2000) conducted video taped experiments and observations on one dog and one owner. They video taped the dog at a window during owners absence and when the owner was returning home. Various observations of the dogs' behavior were conducted in controlled environment and different environments.

These experiments were unique because the tests involved an animal. This presents the possibility that animals and not only humans may have telepathic or ESP abilities.

The test dog was far more at the window waiting when owner was on her way home than when she was not on her way home. The results were displayed in tables and graphs. Replication experiment showed similar results. The suggestion was made that dog was detecting owners' intention to come home. The testing removed the chance that anticipation of the owners return was based on time of day or consistent return times.

Sheldrake and Morgan (2003) conducted trials with one parrot and one owner. One hundred forty-seven, two minute trials, using video tape equipment in the experiment with parrot and owner placed in different rooms. The owner would open up an envelope with a word that the parrot knew how to say. Seventy one trials showed positive results of the parrot saying the word corresponding to the picture that the owner looked at in the other room. . Findings were consistent with the hypothesis that the parrot was responding telepathically to owner's mental activity. This study was unique in that it used a parrot in the experiments.

Sheldrake (2000) noted the most common kinds of seemingly telepathic responses are the anticipation by dogs and cats of their owners coming home. The animals may also anticipate the owner's intention of going away. Also they may anticipate the owner getting ready to feed them. Sheldrake believes that there is much potential for further research on animal telepathy. Sheldrake theorizes that if domestic animals are telepathic with their human owners, then there is a possibility that the animals are telepathic with each other. This may play an important part in the wild. The newsletter discourse sheds some light on telepathy and ESP in the animal world theorizing that telepathy from people to animals usually occurs only when there are close emotional bonds. Emotional bonds may also be an important factor in human telepathy.

Sheldrake, Lawlor, and Turney conducted a telephone survey on perceptive pets. The survey was conducted in London by telephone, with the intent to find out how many pet owners had observed seemingly telepathic abilities in their pets. Fifty-two percent of dog owners claimed that their animals knew in advance when a member of the household was on the way home, compared with twenty-four percent of cat owners. Of the animals that reacted. Twenty-one percent of dogs and nineteen percent of cats were said to do so more than ten minutes before the persons return. Seventy-three percent of dog owners and fifty-two percent of cat owners said their pets knew when the owners were going out before they showed any signs of doing so. Forty-three percent of dog owners and forty-one percent of cat owners said their pets responded to their thoughts or silent com¬mands; and fifty-seven percent of dog owners and thirty-seven percent of cat owners said their pets were sometimes telepathic with them. Forty-six percent of people with pets now and thirty-seven percent of people without pets now said that they had known pets in the past that were telepathic. Thirty nine percent of those with pets now and thirty-eight percent of those currently without pets said they themselves had had psychic experiences. But significantly fewer of those who had never kept pets had had psychic experiences themselves. The results of this survey are compared with two similar surveys in North-West England and in California. The general pattern was remarkably similar in these three very different locations and shows that seemingly telepathic abili¬ties in pets are common. In all locations dogs were more responsive than cats to their owners' thoughts and intentions. Results based on this survey are subject to error since it involves the perceptions of the pet owners without thorough experimental trials to prove claims.

Thomas and Fletcher (2003) conducted test on the mind reading accuracy in intimate relationships. The study tested for the moderating effects of the judge, target, and relationship on mind-reading accuracy during intimate problem-solving interactions. The study used a video-review procedure, multiple perceivers judged multiple targets at different levels of acquaintanceship (dating partners vs. friends vs. strangers). The study also investigated the role of three relationship-level predictors of mind-reading accuracy (for dating couples and friends): relationship satisfaction, closeness, and prior disclosure about the problems discussed."

Their conclusions were that by systematically examining the relationship between the perceiver and the target, the study was able to disentangle the contributions made by the relationship, the target, and the judge in terms of mind-reading accuracy. It was found that all three moderators played a role. It was mentioned that the nature of the relationship between the judge and the target exerted the most profound effects. This moderated the influence of the target and the judge in both expected and unexpected ways.

This was a unique study in that it evaluated intimate relationships in a documented study. An important aspect of this study was the finding that mind-reading accuracy does not reach a plateau at a certain level of friendship, but goes up a notch in the context of intimate romantic relationships. The study does demonstrate acquaintanceship effects in both mind reading and personality domains.

DeGraaf and Houtkooper (2004) evaluated emotional awareness with tests on individuals who have experienced past emotional trauma. Twelve subjects were asked to guess the top down sequence of symbols in an open deck of one hundred Zener Cards. Four cases were studied. The researchers assumption in of this experiment was that a strong unconscious wish or "intentionality" to express the emotions connected with certain pictures could cause relatively more correspondences with Zener cards (ZTs)--or, in fact, their simulations--at certain "sensitive spots. The Zener targets were located at the twelve "sensitive spots," i.e., the exact locations of the twelve pictures. Using the twelve pictures also showed that subjects with the higher trauma scores were able to pick the twelve pictures out and the pictures attracted considerably more displacements of all types than pictures which the subject had left aside in the initial review. The study suggested that these findings could be a result of conscious choice between the pictures mentioned and pictures not mentioned. This may have mirrored the emotional significance that the subjects had also unconsciously bestowed to twelve pictures, which may. This could explain the general lack of significant correlations between trauma scores and displacement types for pictures not mentioned. These tests were a further evaluation in a different way of relating telepathy to emotion.

Murray, Howard, Wilde, Fez, and Simmonds-Moore (2007) tested for telepathy using an immersive virtual environment. They say there are a number of advantages over Ganzfeld work using static or dynamic stimuli or immersive virtual environment. Two hundred males, one hundred twelve females, were tested in pairs at The University of Manchester.

The experimenters used computer technology to set up an immersive virtual reality environment. This was a new type of computerized testing which they felt was superior to the Ganzfeld standard computerized test procedures. The study did not find results to support the Psi hypothesis. The results could be used to argue for the nonexistence of Psi. The Ganzfeld test lasts for two hours and these tests lasted seven minutes. That difference could be criticism for using too short of a time. This too short of test time and test procedures may have resulted in poor results.

Moss and Gengerelli (1967) conducted a controlled experiment on telepathy and emotional stimuli. The study was an attempt to simulate in some small measure the strong affects which appear to accompany spon¬taneous telepathic events. The test used stimuli and after each stimu¬lus episode, the participant spoke their reactions, which were recorded verbatim. After a suitable delay between tests, a second emotional episode was presented.

The study conducted in a controlled laboratory indicates that something like telepathy occurs between two people, isolated from each other, when the Transmitter is emo-tionally aroused and the Receiver is lying down, relaxed. The results showed that seven out of twelve professional psychologists and psychiatrists matched the protocols of fifty experimental T-R teams which is significantly better than that of chance ex-pectation However it was found that under two controlled conditions involving thirteen and ten T-R teams, respectively, only one of the same twelve judges matched in these tests which is better than chance.

This study was included because it was one of the forerunners to current studies on telepathy and emotion. It however showed results that indicated further study was necessary because at that time telepathy was considered abnormal phenomena that most researchers dismissed as coincidence.

Relationship to ESP

Tressold and Prete (2007) studied ESP with subjects under hypnosis. Twelve volunteers (seven males and five females) who attend authors' center were tested. Two types of hypnotic induction were used. Those tested used simple gambling tasks. The author's use of hypnosis to evaluate ESP was compared to other hypnosis studies. The results used standard statistical comparisons. Thirty-three percent of correct hits were recorded in the first session. Twenty-four percent correct hits were recorded in the second session. The results indicated that people in a hypnotic state may be more conducive to psi phenomena. The conclusion was that further study is needed.

Colwell, Schroder, and Sladen (2000) stated that there was evidence to suggest that individuals not only believe in their ability to detect an unseen gaze, but may actually be able to do so. Twelve volunteers were recruited on the basis of their belief in ESP. The volunteers were placed in closed rooms with one way mirrors. The responses were recorded on a computer. The median age was twenty-four with seven men and five women tested who where in the nineteen to forty-nine year old age range. This test was significant because selection of the volunteers were based on their belief in ESP. The test results showed little support for the staring detection effect on non feed back trials. In feedback trials the response bias was present. The two experiments conducted showed no proof that staring could be detected. The authors cited Sheldrake (1994) in this research paper. Sheldrake suggested that the effect is difficult to obtain in artificial conditions, presumably as found in this research study.

Wiseman and Greening (2002) conducted a mass participation experiment where participants were asked to complete an ESP task that involved them guessing the outcome of four random electronic coin tosses. All data was stored on computers. The researchers aimed to help resolve ESP debate by devising a novel procedure for carrying out a large scale force choice ESP Experiment. Mind machine consisted of computer based video clips related to coin tosses. The final database contained twenty seven thousand eight hundred fifty-six participants, Two hundred fifty thousand data points with one hundred thirty-nine thousand forty three data points from five question tests and one hundred ten thousand nine hundred fifty-nine data points from ESP trials. The Over all outcome showed results did not differ from that of chance and all of the internal analyses were non-significant. The mind machine took place in noisy public spaces and not in quiet laboratory surroundings. The study may have failed because forced-choice ESP may not exist.

Owens and Pitman (2004) conducted a study with the aim to

manipulate expectations or attitudes before and during a test of ESP. Thirty one student volunteers at University of Glamorgan were used in tests consisting of six males and twenty five females. They were randomly assigned to four conditions. Participants ages were between eighteen and forty seven with mean of twenty five. Participants were tested individually by one experimenter. Some were given a placebo stating that it enhanced ESP abilities and tested without a placebo. The Experiment used the Australian Sheep-goat scale, a computerized test. Participants asked how they were going to test for ESP before given out a possible Twenty-five correct guesses selected from five possible targets.

Experiment expected to test two connected hypotheses. First that

ESP performance would be positive affected by the manipulation expectance prior and during test and second that this result would be most evident with the indecisive group who did not believe in ESP.

Two Experiments were conducted with both experiments showing that ESP Performance was highest with Placebo or high false expectations. There was a possible relationship to the self-efficacy perspective which increased expectance to the degree of participant success in scoring high in ESP.

Psi Phenomena

Psi denotes anomalous processes of information or energy transfer. These processes may be telepathy or other forms of extrasensory perception. The term is used in relation to all currently unexplained phenomena in physical or biological mechanisms.

Bem and Honorton (retrieved 2009) Does psi exist? According to survey conducted by researchers, most academic psychologist does not think so. One thousand one hundred college professors in the United States surveyed found that fifty-five percent of the natural scientists, sixty-six percent of social scientists and seventy-seven percent of academics in the arts, humanities, and education believe that ESP is either an established fact or a likely possibility. A comparable to psychologist was that only thirty-four percent thought it was possible. An equal number of psychologists believe that ESP is impossibility, a view expressed by only two percent of all other respondents.

The study goes on to list various aspects of areas of psi research.

Alexander and Broughton (2001) developed an automated testing system using equipment called Autoganzfeld H which uses the same hardware and software as that used to control Ganzfeld experiments as that used to accumulate the raw data files by Psychophysical Research Laboratories that closed in 1989. Right brain hemisphere dominance was measured by a procedure called CLB. The results were then related to the Ganzfeld procedure. The scoring was related to the direct hit method. Participant scoring right hemisphere dominance as measured by CLB scored significantly more direct hits in Autoganzfeld than those with left cerebral hemisphere dominance. There was no conclusive evidence of brain and psi phenomena. However, the part of the brain that has potential for further Psi phenomena was identified which makes this study significant.

Carpenter (2008) summarizes research that has been directed toward the Psi phenomena. Carpenter presents information and provides an overview of Psi theory, summary of researchers conducting test, and general findings and hypothesis. There are three pages of references listed in back of research article. This paper is important because it summarizes the theories and beliefs on ESP and Psi, plus some of the research that has been conducted to date. It better helps understand what has been done.

The psi model asserts that the earliest source of potential information comes from our non local engagement with reality. This preconscious process happens very quickly when our conscious experience flows along. Psi is assumed to be quite a normal process, which can quickly be deployed and quickly abandoned. Psi is normally invisible to conscious experience.

Broughton (2006) presents research paper on two stage Psi process. Presentation of information on other aspects of Psi process not previously reported upon. The findings and theories presented are a good general dissemination of information. The two stage Psi model consists of stage one – how ESP 'gets into the system" – which remains a mystery and Stage two which is thought to involve the normal cognitive processes. The research paper implies that if evolution has conferred upon humans the ability to make use of anomalous information then it is likely to follow the pattern in which existing brain systems are adapted and enhanced to confer new advantages. The context of the two-stage model of receptive psi, would lead one to expect that evolution would have adapted existing brain systems to capitalize on anomalous, psi-based information required in order to serve survival goals. The memory systems in the brain have been identified that mediates anomalous information into conscious awareness. However the issue of how the particular memory images are selected remained unaddressed and further research needs to be conducted in this area.

Roe and Holt (2007) conducted a study using forty participants and generated virtual readings consisting of twenty-four statements, eight from each of the three selection methods used in experiments. This was a confirmation study of an earlier study that showed positive results to existence of psi. Results showed significant interaction was found between target lability and sender lability, replicating earlier effect. The study also showed that any PK sender effect is sensitive to situational variables and that these form complex interaction patterns. Researchers stated that goal and process-orientation may be confounded by factors such as activity-passivity that co vary with this variable.

Brain Wave Patterns

Tucker (2009) summarized research just being started by David Poeppel and researchers at The University of California for the US Army. They received a grant to study synthetic telepathy. The study is of electrical impulses in the brain and the impulses effect related to telepathy. They plan to relate brain wave electrical impulses to thought patterns. Researches hope to train subjects to think in code patterns like Morse code with the code being picked up by sensors trained to focus on electromagnetic frequency in the brain and then send detection to a computer or resent to another sensor. Their theory is that motor memory gives a big signal and can be read or extracted. All mental thoughts create electrical signals. This would allow something like helmet to helmet telepathic communication as quoted in the article.

Moulton and Kosslyn (2008) conducted experiments using neuroimaging (MRI) to monitor brain resonance patterns in trying to resolve the Psi debate. The experiments were unique in that they actual monitored brain wave patterns. The findings showed no conclusive evidence and may provide the strongest evidence yet obtained against the existence of paranormal mental phenomena. MRI imaging is hard to read and the Psi influence may be using a different energy field yet undiscovered that relates to anomalous characteristics. MRI's may not accurately pick up the psychic energy field.

Begley (2008) summarizes research studies providing information that could prove the existence of Psi, Telepathy, and ESP are related to brain electrical energy and can be measured with a mind reading dictionary and development of a mind reading machine made possible. Mind reading technology is advancing quickly. The author states that less than three years ago, it was a big deal when studies measured brain activity in people looking at a grating slanted either left or right; fMRI patterns in the visual cortex revealed which grating the volunteers saw.

Research has broken the "content" barrier. Scientists at Carnegie Mellon University showed people drawings of five tools (hammer, drill and the like) and five dwellings (castle, igloo ... ) and asked them to think about each object's properties, uses and any thing else that came to mind. Meanwhile, fMRI measured activity throughout each volunteer's brain. The activity patterns evoked by each object were so distinctive that the computer could tell with seventy-eight percent accuracy when someone was thinking about a hammer and not, say pliers. CMU neuroscientist Marcel Just thinks they can improve the accuracy (which reached ninety-four percent for one person) if people hold still in the fMRI and keep their thoughts from drifting to, say, lunch.

The results have to be replicated by independent labs before they can be accepted. This is the first time any mind-reading technique has achieved such specificity. Remarkably, the activity patterns -- from visual areas to movement area to regions that encode abstract ideas like the feudal associations of a castle -- were eerily similar from one person to another. The report indicated that there is a commonality in how different people's brains represent the same object.

The report indicated that the more detailed the thought is, the more different these patterns get, because different people have different associations for an object or idea The CMU group is determining the brain patterns that encode abstract ideas (honesty, democracy), words and sentences, a big step toward a mind-reading dictionary The article provides the most supportive explanation of the reality of telepathy and ESP related to brain electrical patterns.

Sheldrake (2003) wrote an article explaining his theory on explaining telepathy and other phenomena. Mental fields are rooted in the brains energy fields. The example using magnets can best explain it. Magnetic fields around mag¬nets are rooted in the magnets themselves, or just as the fields of transmission around mobile phones are rooted in the phones and their internal electrical activities. As mag¬netic fields extend around magnets, electromagnetic fields also extend around mobile phones. This analogy can be related to mental fields extending around brains.

Mental fields can be used to help explain telepathy, also the sense of being stared at and other widespread but unex¬plained abilities. Above all, mental fields underlie normal perception. They are an essential part of vision. The mind mental energy related to vision may give of an energy field in the brains processing of this information.

Ganzfeld Experiments

Bem, Palmer, and Broughton (2001) The Ganzfeld database may be a victim of its own success. Ganzfeld experiments have been around for a long time and have become a test standard for ESP testing. Article presents view that the Ganzfeld procedure appears to provide replicable evidence for Psi. The term "psi" denotes anomalous processes of information transfer. Many studies have been conducted and this article lists some of them. The Ganzfeld procedure uses two participants as a sender and as a receiver. It is an established set standard for telepathic studies. The article rated studies that used the Ganzfeld procedure. The researchers suggest that future meta-analyses should distinguish "standard" replications from non-standard extensions of the Ganzfeld procedure test should it become a victim of its own success.

Roe and Holt (2006) used an automated Ganzfeld computer system in their study. Forty trials were conducted. Twenty three trials involved senders and seventeen did not involve senders... The receivers registered a twenty-five percent hit rate. This study indicated that the sender serves some active role in the Ganzfeld experiment ESP sessions. Results varied from a previous study. Generally studies look at consistency of results of the various studies. Twenty-five percent hit rates of receivers are exactly what would be expected by chance. Results were reported in table form.

Roe, Holt, and Simmonds (2003) hypothesized three predictions in the study. They used Ganzfeld GESP protocol .Forty participants using an automated Ganzfeld computer system were tested. The trials used statements and video clips. An attempt was made to distinguish between senders and receivers that participated in the study. The role of sender has been inconsistent in other studies and the relationship of receiver and senders not taken into account. This study used a novel approach for assessing any sender influence. Co variation of performance with the receiver and the sender variables were evaluated. The receivers had a thirty-five percent hit rate which was above the mean chance expectation. The results were listed in table form.

Conclusions

In the literature review on the theme "reality of telepathy", twenty pieces of literature related to the theme were researched. They consisted of telephone surveys, videotaped experiments, experiments on animals, experiments related to emails, evaluation of emotional effects, mothers' interactions with children, and results related to intimate relationships. The large majority of the literature reviewed indicated a greater than chance possibility that telepathy may be involved in the results documented in the literature. Some literature indicated the findings were equal to that of chance; however that literature was in the minority of reports researched.

The literature review on the theme "relationship to ESP" looked at four sources of information regarding this theme. Areas were on hypnotic effects, detecting unseen gazes, the mind machine, and manipulated expectations. Results showed that hypnosis and manipulated expectations could influence and enhance the effects in results searching for evidence of ESP. Computerized experiments may have fell short in getting results because tests were not conducted in a quiet laboratory setting and it did not involve mental interaction with other people, but consisted of trying to guess outcomes.

The theme "Psi Phenomena" had to do with research conducted on anomalous information or unexplained phenomena related to physical or biological mechanisms. Five sources of information were reviewed with one survey showing that the majority of Psychology Professors do not believe in telepathy or ESP. Professors in other fields had a high belief rate in its existence. This survey results may indicate why one of today's most leading researcher in the field is in the biology field and not the psychology field.

This literature review shows some indication that when you remove human interaction in the research study the general results are equal to that of chance.

The literature review has shown that research on the brain is only now getting started. This is a new area of study and the literature review shows research is heading in that exciting direction. There is a good chance of success in developing a device that can read the brains electrical energy fields and can indicate what that individuals mind is thinking. A mind-reading dictionary is being developed that identifies different thoughts that result in brain wave patterns, found to be the same between different individuals. Researchers are also conducting experiments for the military. Sheldrake, a leading researcher theorizes that the mind gives off energy fields that extend around the brain. This energy field can be picked up and read by other peoples' minds. Other research in identifying the characteristics of this energy field needs to be conducted.

Ganzfeld experiments have become a standard of test for telepathy and ESP among Psychologist researchers. The experiment relates to a standard computer test that is used for identifying areas of anomalous interactions. Some experiments show more than chance findings and others show equal to chance findings. Taking the human interaction between individuals out of the research equation and relying upon a computer to give and record response may not be in the right direction to identifying the true nature of telepathy and ESP.

In conclusion, more research should be conducted in areas related to the brains energy fields, efforts to identify this field, plus the mechanism that the mind uses in reading the energy waves given off by a persons' brain need to be conducted. The over all literature review gives an indication that the occurrences of telepathy and ESP may be more than just mere chance. Man is still in the dark ages related to research and understanding of this subject. This literature review compiles and places in one document an overview on the subject and indicates the direction that future research may result in the most success in determining "How real is telepathy and ESP?"

References

Alexander, C. H., & Broughton, R. S. (2001, D). Cerebral Hemisphere Dominance and ESP Performance in the Autoganzfeld. The Journal of Parapsychology, 65(4), 397-416. Retrieved April 4, 2009, from Doane College Wilson web Database.

Begley, S. (2008, Jan 21). Mind Reading is Now Possible. Newsweek, 151(3), 22.

Bem, D. J., Palmer, J., & Broughton, R. S. (2001, S). Updating the Ganzfeld Database: A Victim of Success. The Journal of Parapsychology, 65(3), 207-18. Retrieved April 4, 2009, from Doane College Wilson web Database.

Bem, D. J., & Honorton, C. (n.d.). Does Psi Exist? Replicable Evidence for an Anomalous Process of Information Transfer. Psychological Bulletin, 115(1), 4-18. Retrieved April 4, 2009, from Doane College Ebscohost Database.

Broughton, R. S. (2006, Fall). Memory, Emotion, and the Receptive Psi Process. The Journal of Parapsychology, 70(2), 255-74. Retrieved April 4, 2009, from Doane College Wilson web Database.

Brown, D. J., & Sheldrake, R. (2001, Je). The Anticipation of Telephone Calls: A Survey in California. The Journal of Parapsychology, 65(2), 145-66. Retrieved April 4, 2009, from Doane College Wilson web Database.

Carpenter, J. C. (2008, Spr/Fall). Relations Between ESP and Memory in Light of the First Sight Model of Psi. The Journal of Parapsychology, 72, 47-76. Retrieved April 18, 2009, from Doane College Wilson web Database.

Colwell, J., Schroder, S., & Sladen, D. (2000). The Ability to Detect Unseen Staring: A literature Review and Empirical Tests. The British Journal of

Psychology, 91(1), 71-85. Retrieved April 4, 2009, from Doane College Wilson web Database.

DeGraaf, T. K., & Houtkooper, J. M. (2004, Spr). Anticipatory Awareness of Emotionally Charged Targets by Individuals with Histories of Emotional Trauma. The Journal of Parapsychology, 68(1), 93-127. Retrieved April 4, 2009, from Doane College Wilson web Database.

Moss, T., & Gengerelli, J. A. (1967). Telepathy and Emotional Stimuli. Journal of Abnormal Psychology, 72(4), 341-348. Retrieved April 4, 2009, from Doane College Ebscohost Database.

Moulton, S. T., & Kosslyn, S. M. (2008). Using Neuroimaging to Resolve the Psi Debate. Journal of Cognitive Neuroscience, 20(1), 182-192.

Murray, C. D., Howard, T., Wilde, D., Fox, J., & Simmonds-Moore, C. (2007, Spr/Fall). Testing for Telepathy Using an Immersive Virtual Environment. The Journal of Parapsychology, 71, 105-23. Retrieved April 4, 2009, from Doane College Wilson web Database.

Owens, N . E., & Pitman, J. A. (2004, Spr). The Effect of Manipulating Expectations Both Before and During a Test of ESP. The Journal of Parapsychology, 68(1), 45-63. Retrieved April 4, 2009, from Doane College Wilson web Database.

Polls Show Paranormal Beliefs on the Rise, Evolution Belief on the Decline. (2001). Skeptic, 9(1), 10-11. Retrieved April 4, 2009, from Doane College Wilson web Database.

Roe , C. A., Holt, N. J., & Simmonds, C. A. (2003, Spr). Considering the Sender as a PK Agent in Ganzfeld ESP Studies.

The Journal of Parapsychology, 67(1), 129-45. Retrieved April 4, 2009, from Doane College Wilson web Database.

Roe, C. A., & Holt, N. (2006, Spr). A Further Consideration of the Sender as a PK Agent in Ganzfeld ESP Studies. The Journal of Parapsychology, 69(1), 113-27. Retrieved April 4, 2009, from Doane College Wilson web Database.

Roe, C. A., & Holt, N. J. (2007, Spr). The Effects of Strategy and Feedback on Performance of a PK Task. The Journal of Parapsychology, 70(1), 69-90. Retrieved April 4, 2009, from Doane College Wilson web Database.

Sheldrake, R. (2000, July). The Unexplained Power of Animals. ISAZ Newsletter

Sheldrake, R. (2003). Mind Fields. Resurgence - London - Navern Road

Sheldrake, R. (2005). The Non-Visual Detection of Staring - Response to Commentators. Journal of Consciousness Studies, 12(6), 117-126.

Sheldrake, R. (2006). Apparent Telepathy Between Babies and Nursing Mothers. Sheldrake Papers www.sheldrake.org. Retrieved May 12, 2009, from Google Scholar with keyword telepathy and author Sheldrake Database.

Sheldrake, R., & Morgan, A. (2003). Testing a Language-Using Parrot for Telepathy. Journal of Scientific Exploration, 17(4), 601-616.

Sheldrake, R., & Smart, P. (2000). A Dog that seems to know when his Owner is Coming Home. Journal of Scientific Exploration, 14(2), 223-255.

Sheldrake, R., & Smart, P. (2003, July). Experimental Tests for Telephone Telepathy. Journal of the Society for Psychical Research, 67, 184-199. Retrieved April 4, 2009, from Doane College Ebscohost Database.

Sheldrake, R., & Smart, P. (2003, Spr). Videotaped Experiments on Telephone Telepathy. The Journal of Parapsychology, 67(1), 147-66. Retrieved April 4, 2009, from Doane College Wilson web Database.

Sheldrake, R., & Smart, P. (2005). Testing for Telepathy in Connection with E-mails. Perceptual and Motor Skills, 101, 771-786.

Sheldrake, R., Godwin, H., & Rockell, S. (2004). A Filmed Experiment on Telephone Telepathy with the Nolan Sisters. Journal of the Society for Psychical Research, 68, 168-172.

Sheldrake, R., Lawlor, C., & Turney, J. (n.d.). Perceptive Pets: A Survey in London. Sheldrake Web site www.sheldrake.org.

Smith, J. (2005, S). A New Way of Seeing. The Ecologist, 35(7), 52-6. Retrieved April 18, 2009, from Doane College Wilson web Database.

Thomas, G., & Fletcher, G. J. (2003, Dec). Mind-Reading Accuracy in Intimate Relationships. Journal of Personality and Social Psychology, 85(6), 1079-1094. Retrieved April 4, 2009, from Doane College Ebscohost Database.

To The Editor. (2003, Fall). The Journal of Parapsychology, 67(2), . Retrieved April 4, 2009, from Doane College Wilson web Database.

Tressold, P., & Prete, G. D. (2007, Spr/Fall). ESP under Hypnosis: The

Role of Inductions and Personality Characteristics. The Journal of Parapsychology, 71, 125-37. Retrieved April 4, 2009, from Doane College Wilson web Datab

Tucker, P. (2009, Ja/F). David Poeppel, Master of Synthetic Telepathy. The Futurist, 43(1), 23. Retrieved April 4, 2009, from Doane College Wilson web Database.

Vassy, Z. (2004, Fall). A Study of Telepathy by Classical Conditioning. The Journal of Parapsychology, 68(2), 323-50. Retrieved April 4, 2009, from Doane College Wilson web Database.

Wassner, S. (2005, Mr). Are Twins Mind Readers. National Geographic Kids, (348) pp. 32-3. Retrieved April 18, 2009, from Doane College Wilson web Database.

Wiseman, R., & Greening, E. (2002, N). The Mind Machine: A Mass Participation Experiment into the Possible Existence of Extra-sensory Perception. The British Journal of Psychology, 93(3), 487-99. Retrieved April 4, 2009, from Doane College Wilson web Database.

www.ingramcontent.com/pod-product-compliance
Lightning Source LLC
Chambersburg PA
CBHW040223220526
45473CB00001B/101